Contents

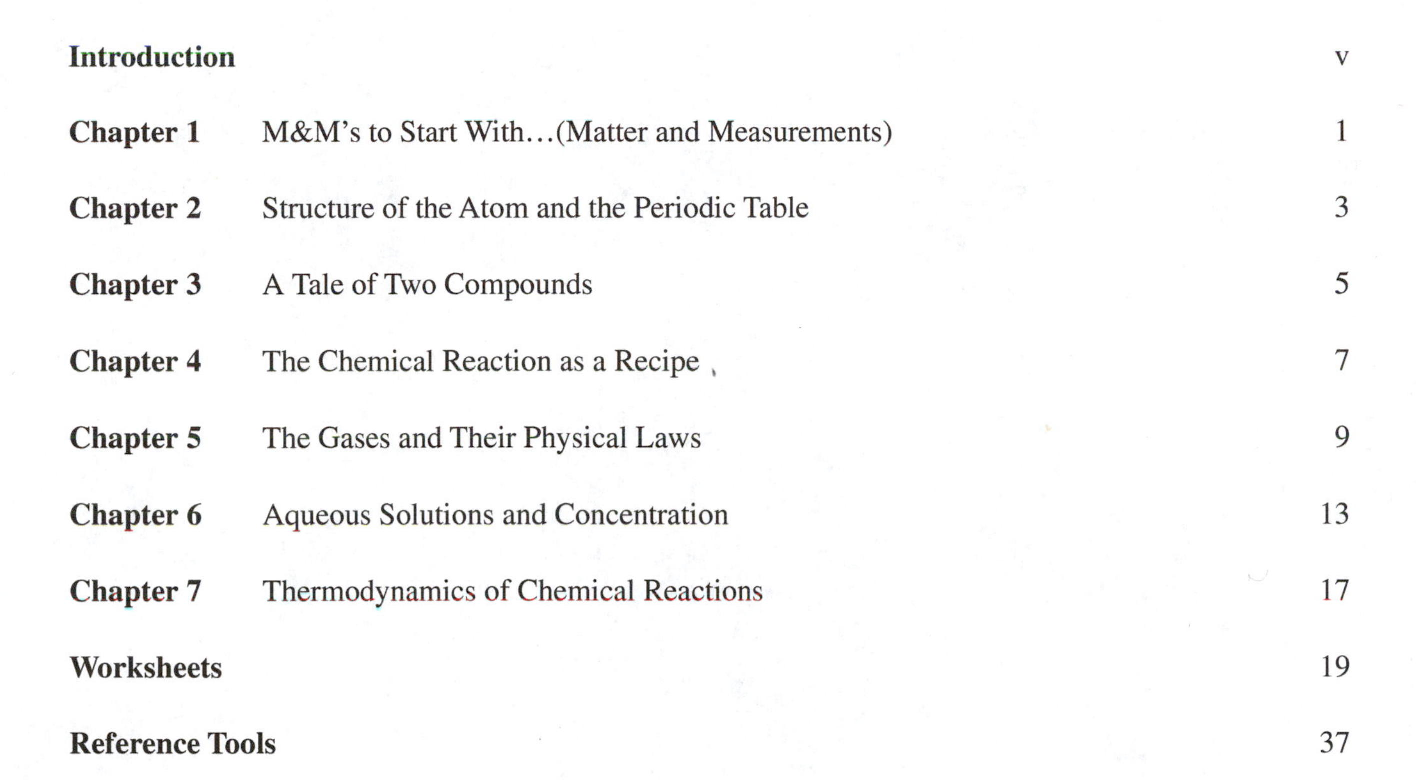

Introduction — To Be or Not To Be...

Although this most famous of lines from Hamlet's soliloquy certainly aims at much deeper philosophical issues than I feel adequately equipped to address, we can nonetheless begin our journey with just such a question, reconfigured to address our concerns when it comes to any discussion of Chemistry. In fact, any of the traditional scholastic sciences (biology, physics, earth sciences, etc.) must first undertake the question of what "IS" and what "ISN'T" available to us to study in a scientific manner.

This leads us directly to defining what is more commonly perceived by us as "stuff," but which we will now call matter.

Words found in any dictionary are typically followed by a letter, or several letters, in parentheses, for example, adj., adv., verb, and so on. Words followed by (*n*), of course, are labeled as "nouns," which most elementary-school lessons tell us is either "a person, place, or thing." We can usually, with some imagination and open-mindedness, picture a noun being held in our hands or glimpsed from near or afar, but either way, something that is directly available to one, or more, of our five senses.

There is, however, an appreciable amount of words listed as nouns in a dictionary that present a bit of a challenge in this area of visualization. A few examples, such as "idea," "opinion," "nightmare," "emotion" are difficult to imagine, precisely, at least from person to person (and even more difficult to truly "sense"). These are, nonetheless important nouns! The progress of human civilization and functioning societies have been founded on such nouns! (Well, maybe we could all do without, or with fewer, nightmares). One difficulty with studying such broad-based "nouns" is that they are neither consistent nor repeatable from person to person, or era to era, and an indeed subject to individual interpretation and critique. Let us leave these, at least for now, to other parts of your college curriculum, such as your psychology, history, and the arts and literature classes.

What this leaves us to study then, are those "nouns" which we can (a) hold in our hands, (b) see with our eyes, (c) taste/smell (when safe!) with our mouths and noses. Right? Well, close, but not exactly…

The study of chemistry and of the ingredients that make up this very large collection of "matter" all around (and including!) us, requires us to actually rely a bit upon our imagination and an ability to draw simplified, but enlarged, replicas of these fundamental bits of matter known as *atoms*.

I have been studying, teaching, and/or practicing chemistry for the better part of my life, but I have never seen an atom. I have never held, nor poked at, "an atom." They are just waaaaay too small for our human senses to discern at that size level. (In one study I was involved with, our team was actually able to "see" a single "molecule" (a collection of attached atoms), but this molecule needed to be rather large and complex and "tagged" with a marker that glowed (or "fluoresced") when viewed using a particular type of microscope affixed with an ultra-fast camera. (So, maybe I should only claim that I have seen a *photograph* of a single molecule).

In our upcoming discussions, therefore, it may prove valuable and helpful to imagine ourselves with the ability to reduce *our* sizes to better discern these atoms and molecules we will encounter. Try imagining, for example, an updated version of "Honey, I Shrunk the Kids…(and they are *still* shrinking)" type of smallness.

Another challenge we will likely face is the idea that, although *all* atoms are extremely small, there also exists a wide range of sizes among the extensive menu of differently "flavored" atoms. For example, the smallest atom, hydrogen, may appear as merely a grain of sand when compared to a different, and relatively speaking, "humongous" atom of barium, which could be imagined as being the size of a bowling ball, by comparison.

Keeping (Complex) Ideas Simple

The idea of "Occam's Razor," theorizes and claims that when faced with a variety of possible explanations to a given question or mystery, the most likely explanation is that which relies upon the least number of variables and the fewest number of underlying, (sometimes unsupported) assumptions or conjectures. Think of these as "leaps-of-faith." Scientists, although not beholden to this theory of simplified explanation, prefer to at least use the idea of Occam's Razor as a starting point in their investigations (as did Sherlock Holmes!). Using such a methodology, scientists are able to not only often provide satisfactory (if only preliminary) answers to the questions posed, but also that this simplified approach makes repeating such experiments or investigations by other scientists that much easier. (This "repeatability of experimental methodology" by other scientists, is not only promoted, but mandatory in the ongoing discovery process of science.)

I only tell you this to prepare you…

The title of this course is "Introductory Chemistry."

Not "Organic" Chemistry, not "Physical" Chemistry, not "Analytical," or even "General" Chemistry.

Introductory Chemistry: Our discussions in my classroom, upon which this textbook is based, are kept at a level that best aligns with Occam's Razor, or as is more contemporarily known, "K.I.S.S.—Keep It Simple, _____." The depth of this course may only amount to a surface skim across a kiddie pool, but as past students have attested, if diligently pursued and practiced, it will prepare, among others:

Those with little to no background in Chemistry specifically, or science studies, in general.

Those seeking a preparatory course in the ABC's of Chemistry, prior to taking the deeper, higher level courses (see above).

Those preparing to apply to an Allied Health or medical program.

Those embarking on an elementary/high-school teaching career.

Those with an interest in Occam's Razor and the KISS approach to discovery.

Personally, I find this course to be a chance to not only help students gain a comfortable, yet firm, grasp on the fundamentals of chemistry, but also my opportunity to pay homage to many of the great science practitioners, and perhaps equally important, science *communicators*, such as Drs. Carl Sagan, Richard Feynman, and Neil DeGrasse-Tyson.

Herein, together, we shall:

Re-enact the classic TV sitcoms/dramas involving partners of policemen in squad cars.

- Re-build Humpty Dumpty (without waiting on all the King's Men).
- Get a little crazy with balloons and (not as crazy with) thermonuclear devices.
- Meet the tiniest nightclub bouncer ever.
- Get to play the roles of both an electrifying Cupid and gruff construction foreman as we create chemical compounds.
- Journey down the Yellow Brick Road of the Periodic Table, keeping a travel diary (or "blog," I suppose) of our journey and the observations we make along the way.
- Several other corny, yet hopefully helpful, anecdotal sojourns down to "Imagination Station."

And finally, I must admit before we begin, that a few of the explanations I will attempt to provide are likely to not be as "simple" in reality as they may seem here. Those that continue to higher level studies in chemistry will hopefully better appreciate my attempt(s) at simplification, whence they realize the complex nature of orbital hybridizations, enhanced electron affinities, or even quantum mechanical concepts (I am still working on *that* one, 20 years after graduate school, where I liked the course so much that I took it three times!)

Chapter 1

M&M's to Start With (Matter and Measurements)

As we discussed earlier, we ought to limit our scope here to the study of "stuff" that meets a certain physical, or "sense-able" level. In order for something to qualify as matter, therefore, a two-pronged approach is best. We can safely assume that anything that has (and that can be measured!) *mass* and *volume* will suffice to qualify as *matter*.

Mass

Sometimes the simplest concepts, those which we more often than not take for granted, have the deepest explanations. Mass is just one such concept. Let us just get this out of the way…If you have ever stepped on a bathroom scale, place fruit upon the grocery store scale, or at the airport tossed your luggage up on a metal platform to be measured, then you already know enough about "mass" to get by just fine. Do notrepeat this in Physics class (mine or anyone else's), but since, for the most part, we will be confining our discussions here to planet Earth, the distinction is subtle, and can be simplified. Think of a sample of matter's mass as being how much it weighs. ←(not completely true, but OK for here).

Now when measuring mass, or any other property, it is not enough to simply state a number (2, 45, 0.34, etc.). We must always report both a numerical value AND an appropriate unit (or word).

For mass, the common units (at least in the United States) include pound, ounce, and ton. The basic unit of mass in chemistry is the gram. A gram is approximately the mass of one standard office paperclip. Not really large, but easily measured on the scales typically used in a chemistry setting. If need be, we can deploy the use of the Metric Prefixes to collectively measure 1,000 g at a time as 1 kg (see Appendix B for a Metric System tutorial).

Here are some examples of mass measurements being made:

1 standard paperclip = 1 g

1 pet kitten = 1,355 g (or 1.355 <u>k</u>g)

1 "dash" of table salt = 0.19 g (or 190 mg)

The bottom line(s) here are that:

- Mass is a fundamental property of matter, and is easily measured directly (if we have a usable scale).
- The "grams" unit is the preferred, and in fact, "miraculously" (more on this later) appropriate scale for most laboratory measurements in chemistry.

The other required property that something must have to qualify as "matter" is volume.

Volume

Most objects or substances that we encounter each day possess a degree of shape to them. Sure, clouds change shape constantly, as can running water, but again most objects, such as a car (kinda boxy, sitting upon roundish rubber discs), or a pencil (long and slender, pointed at one end, pliable at the other) have some sort of describable shape to them. In recognizing the shape of something, we are in fact acknowledging that the "something" takes up (or "occupies") space, and therefore has a measurable property of volume.

The volume of a regularly shaped object, like a cube, a cone, a sphere can be calculated usually using some formula from geometry that you may (or may not) recall ("pies are squared? Or cubed?"). Fortunately, there exists a simpler way of measuring an object's volume that does not require the use of algebraic formulas.

This is known as the water-displacement technique, and no doubt you have used its principle at some point. When we draw (fill) a bath for ourselves, we typically, if we are watching carefully do not allow the tub to fill much past half-way. Why is this? Well, of course, we realize (maybe from earlier, sloshy attempts) that when we place ourselves in the tub, the water level will rise, even though we have not added any additional water. If our bathtub has little markings for volume (most do not, of course), then we could record the level of water both before, and then after, we get in the tub. By subtracting one number from the other, we can somewhat accurately measure our own body's volume! (*Note*: We would actually need to completely submerge ourselves, holding our breath, and likely require the assistance of someone else to make the second measurement accurately).

Chapter 2

Structure of the Atom and the Periodic Table

In this chapter, we will first discuss some general properties that all atoms of each and every element share. We will meet a family of subatomic particles, three different species in all, each with its own set of "responsibilities" to the atom which it comprises. Luckily, two of these family members, although functionally important, are rather easy to investigate as they are always found in the same region of *any* atom. The third member of this family of subatomic particles is much more active and tougher to pin down. In fact, members of this third type of particle are often discarded by atoms, while at other times, they are recruited to (or rejoin!) an atom. These particles can actually even be claimed, or shared, between two different atoms!

Once we are familiar with these basic structural features that comprise any atom, we will begin to look more closely at what makes atoms of different elements unique from others, and what their position on the Periodic Table tells us about them. The more familiar we become with using, as a tool, the Periodic Table, the easier certain tendencies arising later will be to identify.

General Features of Any Atom

Regardless of which element you choose to examine from the Periodic Table, it will share two main features with all other atoms of different elements as well. If you were to draw, as well as you can, a circle approximately the same size as a quarter or half-dollar, you have drawn what is called the *nucleus* of the atom. The term "nucleus" has several uses throughout science, but in chemistry, we may best imagine the nucleus as the dense, solid, "heavy" (relatively speaking) core of any atom...kind of like the pit found at the center of a peach!

Although the nucleus of an atom does not technically consist of an outer shell, it may be helpful to imagine it as such, for the time being. And, while we are at it, let us imagine this nuclear shell as also being *transparent*, to our eyes at least, so we can glimpse inside easily. We will return to the nucleus, and its contents, in a moment.

The other feature that *all* atoms have in common is a zone outside of the nucleus, extending outward spherically. Think of this zone as perhaps resembling the earth's atmosphere, where although there is a whole lot of available space, aside from some birds, kites, clouds, and airplanes, there is not a much else that we can see in our atmosphere. Such is the same in this region of the atom. This region is where we will expect to find the third subatomic particle mentioned above, the electrons. Therefore, most chemists refer to this "atmospheric," nearly empty, region of the atom as the "electron cloud region", or just "the electron cloud."

The Nucleus and Its Contents

Protons

Back to the core nucleus again. Inside, we find two different types of subatomic particles. They are a little difficult to differentiate from each other, as the two types appear to be approximately the same mass and volume. Also, although they seem to be very tightly crammed into such a small space, they are able to jostle and move around to some extent. The first type of particle we will describe is the proton. Each proton inside the nucleus (and there are several that we can see from where we are observing) tend to keep as far apart from each other as they possibly can. This, of course, is because each and every proton, by definition, is a positively (+) charged particle, and particles that are all positive (or all negative) tend to not want to be near each other. You may have heard this as the tail-end of the statement, opposite charges are attracted to each other, but **similar charges repel each other**.

Protons provide both mass to the atom's nucleus, (more on this later), but more importantly, the protons literally serve as the "identification" for any atom. The number of protons that we find inside the nucleus tells us which element from the Periodic Table we are investigating or using! And the number of protons in each atom's nucleus is simply the element's numerical address on the Periodic Table. So, imagine that the nucleus were looking at contains a total of three protons. The three protons inside the nucleus identify this atom, without a doubt, as an atom of the element "lithium." Lithium is found on every Periodic Table in Box #3 (2nd row, 1st column). If we found an atom that was found to have 87 protons packed inside of its nucleus, we would undoubtedly be looking at an atom of the element "francium, Fr." There are 13 protons inside the nucleus of every "aluminum, Al" atom, 40 protons in the nucleus of any "argon, Ar" atom, and so on. The protons do not really have much more use other than providing mass and identification, but this should not be dismissed as unimportant. What makes the precious metal element "gold, Au" different from the highly toxic element "mercury, Hg" is merely one proton.

Neutrons

The other particle that we find inside the nucleus is the neutron. Neutrons are often described as simply "extra baggage" or "dead-weight," as they themselves have neither a positive, nor a negative-charge associated with them. They are quite literally, "neutral." Oh, how dull and boring you may think. But, ahhh, allow me to hype my favorite of all the subatomic particles, and try to convince you otherwise.

Neutrons, having no charge, are not influenced, nor affected in any way whatsoever by the protons. In fact, their neutrality seems to be the neutrons' most powerful tool! See, neutrons are not impressed by the (+) charge on the protons, not at all. In fact, it is the neutrons that we might wish to truly thank for keeping everything from falling apart, or from having had already exploded. Let me explain further.

You see, the experience of being a proton inside of a nucleus is similar to that of maybe being one of three triplet sisters, all having to share the same mirror for their entire lives. In other words, the protons would like some space between each other, please! As such, the protons, during their entire existence inside the nucleus, (which may be forever!) are struggling to escape to freedom (and elbow-room/privacy) from the other protons. And, they would, were it not for our heroes, the neutrons.

You may prefer to imagine the neutrons' role within the nucleus as being that of packing material (like those Styrofoam peanuts), or as reverse nightclub bouncers (they keep you IN, and prevent you from leaving!)

Soon, we will begin to discuss the various ways in which atoms can combine with other atoms through chemical bonding to form chemical compounds. Regardless of the method of combination, the compounds are formulated through the interaction of the most "active" members of the subatomic particle family, the electrons.

Chapter 3

A Tale of Two Compounds

Okay, now that we have been introduced to the "classic menu" of nature in the form of the Periodic Table and all of the flavors of elements it arranges for us, it is time to look at the next level (or class) of matter, namely chemical compounds.

Think of a chemical compound as a combination of several different flavors of ice cream served in a sundae dish at your favorite Olde Tyme Ice Cream Shoppe. First of all, in any respectable Ice Cream shop, there are literally millions of different two–three or more scoops combinations of ice cream sundaes. But perhaps, more importantly, it has been shown that not all combinations are possible, for a variety of reasons, several of which we shall discuss as follows.

The "Flavors"

So, we spent a good amount of time discussing the structure of atoms, including what they all share in common, and more importantly what it is that distinguishes one element (or "flavor") of atom from another.

With somewhere near 115 possible elements to choose from, it would be helpful perhaps to eliminate a few of them from discussion completely! For reasons we discussed earlier, the Noble Gases of Group 18 are well known for *not* combining with any other element to form compounds. The fact that each member of the Noble Gas group has a completely, and perfectly filled valence shell of electrons (typically eight of them), they have no inclination to either gain or lose any of these electrons. They are fine "as is," thank you very much.

Also, we will go ahead any eliminate, except on rare occasions, the f-block elements, found at the bottom of most Periodic Tables. Many of these elements are actually artificial, and their "lifetimes" are typically very, very short, choosing rather to decay back into small, more familiar elements.

But, that still leaves us with an appreciably large roster of elements that can potentially combine into chemical compounds.

Note: Regardless, of which type of chemical compound we are discussing, it is worth remembering that any atoms involved in the formation of the new chemical compound retain their identity, as you will recall, it is the number of protons inside an atom's nucleus that identifies it. Whenever we discuss the existence of chemical compounds created through the formation of chemical attachments, or "bonds," we must realize that it is only the outer, electron-cloud, portion of any atom that is involved.

Chapter 4

The Chemical Reaction as a Recipe

The heart of any discussion involving Chemistry typically involves a reference to *change*. "Change" here refers to the interaction of two or more chemically unique substances, reacting with one another to produce completely different substances as products.

Of course, the options for chemical reaction ingredients can potentially be any element, (mono- or di-atomic) as well as any chemical compound, and the same can be said for the combination of chemical produced that result.

The general form of any chemical reaction, therefore, can be summarized as:

$$\text{Reactant(s)} \rightarrow \text{Product(s)}$$

However, this simplified representation is about as specific a statement as "Groceries → Dinner." There are a great many forms of chemical reactions, some that involve a tremendous number of different reactants, others that form a large variety of different products. Below, however, are the five most common types of chemical reactions.

Five Common Types of Chemical Reactions:

	Reactant(s)		**Product(s)**
Synthesis:	A + B + C	$\rightarrow$	ABC

The key to recognizing a synthesis type of reaction is noticing that there is always one, and only one chemical *product*. In the example above, three reactants (A, B, and C) combine to form the product with the formula "ABC."

Example: $K + Cl_2 + O_2 \rightarrow KClO_3$

	Reactant(s)		**Product(s)**
Decomposition:	ABCD	$\rightarrow$	A + BC + D

The key to recognizing a decomposition type of reaction is noticing that there is always one, and only one chemical *reactant*. In the example above, one, lone reactant (ABCD) breaks-down/decomposes to form several products (A, BC, and D).

Example: $Mg_3(PO_4)_2 \rightarrow Mg + PO_3 + O_2$

	Reactant(s)		**Product(s)**
Combustion:	$C_xH_y + O_2$	$\rightarrow$	$CO_2 + H_2O$
	$C_xH_yO_z + O_2$	$\rightarrow$	$CO_2 + H_2O$

A combustion reaction is perhaps the most recognizable type of reaction, as the products are always the same, namely CO_2 and H_2O. Additionally, as the name "combustion" implies, these reactions requiring O_2 (oxygen gas) as a reactant (otherwise, the combusting fire is extinguished).

And finally, combustion reactions are recognizable based on the reactant that is typically shown first, the C_xH_y (hydrocarbon) compound, or the $C_xH_yO_z$ (carbohydrate) compound.

Examples: $C_3H_8 + O_2 \rightarrow CO_2 + H_2O$
(propane)

$C_6H_{10}O_5 + O_2 \rightarrow CO_2 + H_2O$
wood/(cellulose)

	Reactant(s)		**Product(s)**
Single Replacement:	AB + C	$\rightarrow$	A + BC

In a single replacement reaction, there are two reactants. One reactant is a binary (2-element) compound, and the other is a lone element. The products here are a different binary (2-element) compound and a different isolated/lone element.

Examples: $Mg + HBr \rightarrow MgBr_2 + H_2$

$Fe_3P_2 + O_2 \rightarrow FeO + P$

	Reactant(s)		**Product(s)**
Double Replacement:	AB + CD	$\rightarrow$	AD + C

A double replacement reactions involves two ionic compounds reacting to form two different ionic compound products, simply by "swapping" their respective (±) ionic partners.

Examples: $Na_2O + AlBr_3 \rightarrow NaBr + Al_2O_3$

$(NH_4)_3PO_4 + Ag_2SO_4 \rightarrow (NH_4)_2SO_4 + Ag_3PO_4$

These will cover the majority of the types of chemical reactions you will encounter in this and other Introductory Chemistry classes.

Balancing Chemical Reactions—Getting The Recipe Amounts Correct

You may have noticed, as we introduced the five reaction types above that, in most cases, although the same elements appear on both the reactant and product sides of each reaction, that the number of atoms for each element is not always, in fact rarely, equal on both sides of the reaction. This is not unusual at all, but does indicate that we must be sure to properly balance the chemical reaction before we are able to utilize the chemical reaction as a "recipe."

Make an identical list, on each side of the reaction, of each and every element you see.

Now, carefully count how many atoms of each element you have on the reactant side, and then do a simple count for each element on the product side. (Be careful to "multiply-by-distribution" any elements that are found contained with a set of parentheses.)

Chapter 5

The Gases and Their Physical Laws

In our day-to-day lives, we constantly encounter the objects around us, be it the softness of the bed from which you rise, to the cement of the sidewalk along your street. The steaming liquid water infused with molecules of caffeine (coffee) is or flavonoids (tea) help propel many of us through our day. These daily encounters with these solid and liquid substances are as common to us as the air we breathe (foreshadowing).

But, how often do we acknowledge that in between the coffee refills and all the while as we sleep in our beds, we are utterly dependent upon another form, or state, or matter. The air that we breathe is a perpetual requirement, but is often the least appreciated, much less, understood examples of matter that we are constantly in contact with.

What we call "air," is of course neither a solid nor a liquid, by definition, although "air" allows both solids and liquids to pass through, and/or be carried along by, it. Instead, the air that we breathe is a *gas*. Actually, it is known as a *gas solution*, as it typically consists as a homogeneous mixture of at least five or six different chemical compounds, in specific proportions, randomly mixed together, and each of these molecules behaving as if it were the only molecule around!

Gases are:

1. Unique!—They are the only familiar state of matter wherein the molecules act independently and are essentially unaffected by the presence of any other gas molecule…
2. Powerful!—At the onset of the Industrial Revolution, scientists realized that gases represent a high energy state of matter, and that the energy of a gas (specifically, the kinetic energy) could be harnessed into performing work for us…
3. Highly "adjustable"—Gases can be heated and cooled, just like any solid or liquid can be, but gases respond to temperature changes with dramatic changes in the pressure they exert and the amount of space, or volume, that they occupy…

The Gas "Control Panel"

Whenever we discuss a sample of gas, it may be helpful to imagine the gas contained within some sort of flexible balloon. This visualization helps us to isolate the gas under study and sets up very clear demarcations between what *is* the gas and what *isn't*. (Of course, this is one of the great challenges in meteorology, the study of the mechanics of the Earth's atmosphere, which is, of course just one big gas sample!)

In studying a gas, even when first encountering a gas, there are four characteristics that we can measure and determine at that one given moment of time. However, be careful. When measuring one particular characteristic of a gas, it is frustratingly easy to inadvertently affect one or more of the others as well. Here are the four variables (things that can change easily…) for any confined gas sample:

Pressure (P): A measure of the static force per unit area that a gas experiences and exerts. The units used for pressure are various, but we will typically use the "atmosphere" or "atm." This is a convenient unit to use, as it describes the standard pressure that the Earth's atmosphere experiences on a normal day at sea level. 1.0 atm is defined as Standard Pressure. [Other units of pressure you may encounter, or even be familiar with include pounds per square inch (psi, lb/in^2), millibars, Pascals, and millimeters of mercury liquid (mmHg)].

Volume (V): This one's easy, as we are simply measuring the amount of space that the confined gas occupies. The proper unit to use for volume is the liter, so be sure to convert any volume measurements, from, say, milliliters or centiliters, to regular liters.

Temperature (T): Measuring temperature is likely very familiar to most of us, whether it be the indoor or outdoor temperature of the gas sample, the temperature inside an oven, or our own body temperature. Here, though, is where the familiarity may end.

Here in the United States, we still use the Fahrenheit scale to measure temperature. In most other parts of the world, the Celsius scale is used to measure temperature. Regardless of which scale is used around the world, both the Fahrenheit and Celsius scales reflect the possibility of negative temperatures (below "0"). Since, we will be using values for temperature in some equations and formulas soon, it is important that we use a scale that is designed *not* to reflect any negative temperatures. As it turns out, many formulas in science require what is known as an "absolute" temperature, a value that is *always* greater than "0."

This particular temperature scale is known as the Kelvin scale.

The somewhat cumbersome temperature conversion formulas for Fahrenheit ←→ Celsius can be found in Appendix A, but the conversion from Celsius to Kelvin is very simple, since the Kelvin scale is essentially the same as the Celsius scale, each degree on both scales being of equal size.

$$\text{Kelvin} = \text{Celsius} + 273$$

This, indirectly, tells us that there does exist a theoretical *lowest* temperature possible, namely 0 degrees Kelvin ("0" K), or −273 degrees Celsius (−273 °C). This temperature, never actually reached by human experimentation, is known as absolute zero. Things can get a bit strange down at such cold temperatures, as all kinetic energy is removed, and by definition (of temperature), the object in question (an atom, a molecule, a bowling ball) would have zero energy!

For our purposes, just remember to *always* convert Celsius temperature to Kelvin temperatures before you do any calculations using the Gas laws.

Moles (n): We learned earlier how each and every element and chemical compound has its own intrinsic molar mass. Also, we have seen that chemical reactions are always written with *coefficients* indicating the number of moles of each reactant and product, not *grams*. Similarly, although we can actually measure the mass any sample of gas in grams, the relationships we will be looking at require us to convert grams → moles prior to using any of the Gas law formulas.

So, to review, we will be working with gas samples that have the measureable characteristics listed below, but must always be discussed in the associated units (which may, at times, require conversion activity before solving any given problem.

And, finally, before we begin looking at adjusting individual Control Panel dials, and observing the effect the movement of each has on the others, here are a couple of somewhat useful tidbits of information to keep handy when working with gas sample calculations:

1. Oftentimes, gas sample experiments are set-up under predetermined, agreed-upon, conditions known as Standard Temperature and Pressure (STP). These three letters tell us two immediate pieces of information, and should be incorporated into your "Decoding List."

Specifically, STP means:

Pressure (P) = 1.0 atm

Temperature (T) = 273 K (or 0 °C…brrrr!)

2. There is a very reliable "rule of thumb" when it comes to STP conditions, but it must be applied carefully. It describes the volume that one mole of *any* gas has under STP conditions. (The values for moles and volume can be adjusted, stoichiometrically, as shown below.)
3. Gas law "Rule of Thumb": One mole of *any* gas, at STP conditions will occupy a volume of 22.4 L.
4. Similarly,

   two moles of *any* gas, at STP conditions will occupy a volume of 44.8 L.

   and

   one-half mole of *any* gas, at STP conditions will occupy a volume of 11.2 L.

This shortcut will be much more useful in the Gas laws (Part II)—"snapshot" laws toward the end of the following section.

The Gas Laws—Part I—"Before-and-After" Laws

Instead of introducing the first set of Gas laws one at a time, below you will find a set of four "before-and-after" Gas laws. These laws describe how a gas behaves when certain combination of dials on the Control Panel are moved. Notice that, with one exception, in each of these laws, there are only two dials being adjusted.

$$\text{Boyle's Law: } P_1V_1 = P_2V_2$$

$$\text{Charles' Law: } \frac{V_1}{T_1} = \frac{V_2}{T_2}$$

$$\text{Avogadro's Law: } \frac{V_1}{n_1} = \frac{V_2}{n_2}$$

$$\text{The Combined Gas Law: } \frac{P_1V_1}{T_1} = \frac{P_2V_2}{T_2}$$

Note: The small subscripts of "1" (before) and "2" (after) are simply the method I personally use. Feel free to use any subscripts you would prefer, like "x" for "before" and "y" for "after," or "I" for initial and "f" for "final." Make sure that you are comfortable with whichever system you choose, as we will be performing some algebraic rearrangements to these formulas, and it is important to keep track of the terms as they are reworked and moved around.

Worksheet Packet (Before/After Laws)

The Gas Laws—Part II ("Snapshot" Laws)

There are also Gas laws that can be used to determine a piece of "missing info" form the Gas Control Panel, (*P V N T*). These laws are typically not used for measuring any change occurring, but rather to fully define a particular gas sample at one moment in time, or provide a complete "snapshot" of the gas.

The first "snapshot" Gas law we will look at is known as Dalton's law. This is perhaps the most convenient and easy-to-solve Gas laws to solve, as it deals with only one dial on the Control Panel, namely, pressure (*P*).

Dalton's law states, that in any homogeneously mixed gas mixture, the overall or Total Pressure of the gas mixture is equal to the sum of the individual pressures of the various gases comprising the sample, or:

$$\text{Dalton's law: } P_T = p_1 + p_2 + p_3 + p_4 + p_5 + \ldots.$$

The other "snapshot" Gas law is called the Ideal Gas Law. The name "Ideal" likely came from the hypothesis that, in fact, the product of a gas's pressure (P) and volume (V) is equal to the product of the gas's temperature (T) and the # of moles of the gas (n), or

$$PV = nT$$

But…this initial attempt was found to be flawed…

This first-attempt form of the Ideal Gas Law was found to be true at conditions similar to those found in our own atmosphere's habitable regions (i.e., sea level), but did not predict accurate results at higher temperatures and pressures, where it appeared that the normally nonreactive gas molecules would actually begin to chemically react with each other from time to time.

So, the "Ideal" Gas Law needed a correction term to account for these chemical interactions at higher temperatures and pressures. The value that found to correct the Ideal Gas Law is now called "R," and is known as the gas-law constant. It has no true scientific meaning, and merely comports with physical observations by experiment. Therefore, the value and unit for "R" looks a bit intimidating.

$$R = 0.0821 \text{ L atm/mole K}$$

and the Ideal Gas Law, in its corrected form, is now known as $PV = nRT$.

Although the Ideal Gas Law is easy to commit to memory be reciting "piv – nert", in its original $PV = nRT$ form, it is *not* solved or rearranged for any single variable. This law is useful for finding the missing piece of information for the "snapshot" moment for the gas, but it must be practiced with in order to do so correctly.

Also, it is absolutely critical that, when using the Ideal Gas Law, the standard units for each variable are used. Otherwise, the "R" value will not be valid. A reminder:

P—atm

V—Liters

T—Kelvin

n—moles

Chapter 6

Aqueous Solutions and Concentration

Review—Mixtures: Heterogeneous or Homogeneous

Chemical Compound Types: Ionic or Covalent

Earlier, when discussing mixtures, we noted the distinguishing characteristic of a heterogeneous mixture (looks different throughout; a supreme pizza, or garden salad) and a homogeneous mixture (looks uniform and consistent throughout; e.g., milk, Kool-Aid, coffee, tea).

We also made note of the fact that, very often, the appearance of a homogeneous mixture can be misleading and may lead us to incorrectly characterize such a mixture to be a pure compound or even a pure element. In fact, it is often the case that the *only* person completely sure of the mixture's composition is the person that prepared it in the first place! (Another excellent reason to *always* label the contents of any prepared solution, so as to prevent mistaking it for something that it is not.)

Without citing any statistics or data, it is still reasonable to say that nearly *all* of the substances which we interact on a daily basis are, in fact, mixtures, with most of them being homogeneous and consistent in appearance unless they have been altered with writing, paint, Magic Marker, and so on, after the fact. Think of the vast amounts of plastics, rubber, ceramic material, glass, and so on that surrounds us every day. All of these are examples of uniformly blended, specifically molded and contoured, homogeneous mixtures, most of which are solids, but likely contain a vast list of "ingredients."

The fact is that pure elements or compounds are rather rare in their occurrence in our everyday lives. And these mixtures need not always be in the solid, easily recognizable state with which we are so familiar. Even the common air that we all breathe is a homogeneous mixture … a combination of N_2, O_2, CO_2, H_2O molecules, Ar atoms. (Unfortunately, more and more, there are other ingredients in our air that are less than "natural," including CO, O_3, NO_x, and SO_x, where $x = 1, 2,$ or 3 atoms of oxygen.)

However, without a doubt, the most common and utile type of homogeneous mixture is the aqueous solution. These types of homogenous mixtures, where something (usually solid, but not always), is uniformly dissolved in water, often (but certainly not always) forming a colored solution.

Before we go much further, a few definitions applicable to our discussion of solutions.

Definitions

Solution: A homogeneous mixture of two or more substances. The homogeneous mixture may be of the solid, gas, or cost commonly liquid state.

Solvent: The portion of a homogenous mixture in *greatest* amount, both in mass and volume. The solvent is the portion of a solution that we can think of as "doing the dissolving."

Solute: The material that is added to a solvent and dissolved completely. This is the portion of a solution that "gets dissolved." Most commonly, this is a powdered or granular ionic compound (salt), but anything else (gases, other liquids) that will dissolve in a given solvent can be considered a solute.

Soluble: Generally meant to describe a particular solute as being "dissolvable" in water (or another solvent). In our examples, we will assume that all of our ionic compound solutes are soluble in water (although, there exists certain important exceptions … Ever passed a kidney stone?)

Insoluble: Indicates a substance that is unable to be dissolved by a given solvent, either due to observed solvation "rules" or due to the fact that the given amount of solvent already has dissolved as much as it can, and therefore, the solute is not able to be dissolved (unless, perhaps, more solvent is added!)

Saturated: A solution is described as "saturated" when a no additional solute is able to be dissolved. Any given volume of solvent has a finite (limited) amount of solute that it can dissolve.

Unsaturated: A solution is described as unsaturated if the amount of solvent is able to still dissolve added solute. This can be tricky to determine experimentally, since once you see solute no longer dissolving (as it settles on the bottom), well, you have reached the point at which the solution is saturated.

Types of Solutions (The "Usual," and The Not-So-Usual...)

Very soon, as we begin to characterize a solution by its strength (or concentration), we will typically be referring to a solution in which a powdered or crystalline solute is dissolved in the most universal and common of all solutes, H_2O. These types of solutions, where water is the solvent are known as aqueous solutions (the solute, in these cases, may be liquid or even gas in nature, but is typically some sort of powdered solid crystal or salt).

However, since by definition, a solution is simply:

"…a homogeneous mixture of two (or more) substances…"

We will spend at least some amount of time, in class, discussing other types of solutions found in either the solid state or the gas state.

However, when we typically discuss solutions in chemistry and tin the allied health professions, we are certainly speaking of a water-based aqueous solution. H_2O is not only the most common solvent used in solutions because of its availability, but also because the H_2O molecule is custom-built to dissolve the widest range of any other liquid solvent. Water is therefore sometimes called the "universal solvent." (More on this later on…)

Measuring Solution Concentrations

Concentration, as a general term, can be thought of as how "crowded" or "packed-full" something is, and can be described, (again, generally) as:

$$\text{Concentration} = \text{amount of solute/amount of solvent}$$

The term "amount" here is not specific enough, so in chemistry we use a particular type of concentration measurement known as molarity. As the name implies, the concept of "mole" plays a large part here as well. Molarity (M), in formula form, is defined as:

$$M = \text{moles of solute/Liters of solvent (water)}$$

or, more succinctly, $M = n/L$

As with any mathematical formula, it is worth taking some time to practice rearranging the formula to solve for the different variables.

Form 1	Form 2	Form 3
$M = n/L$	$n = ?$	$L = ?$

The worksheet practice problems (and real life!) will, at different times, require you to employ each of these versions of the molarity formula. It is worth practicing with these rearrangements.

Keep in mind that the "*n*" (moles) and "*L*" (liters) variables *must* be converted to moles and Liters, respectively, *before* using them in any manner with this formula.

Dilutions

The term "dilution" typically refers to the "watering-down" of something, or perhaps making something less complicated and simpler. This is true, as well, in chemistry when using the dilution formula below, and implies that there is yet again a "before/after" situation set-up similar to what we saw with the Gas laws (Part I).

The dilution formula, in fact, more than closely resembles Boyle's law which described the effect that pressure (*P*) has on volume (*V*), specifically that $P_1V_1 = P_2V_2$.

In the dilution formula, the "*P*" terms is simply replaced with the molarity or "*M*" term.

Dilution formula: $\mathbf{M_1V_1 = M_2V_2}$

Once again, we can expect to observe an inversely proportional relationship here, namely that as volume (*V*) is increased (addition of more H_2O in our aqueous cases), that the value for molarity should decrease. Of course, this is in fact what we observe anytime that we add water to some very tart or strong beverage, like cranberry juice.

Conversely, although less commonly, if the volume of the solvent were to decrease (through evaporation), then the molarity will increase. (Same amount of solute, but now being held by a smaller amount of solvent.)

Chapter 7

Thermodynamics of Chemical Reactions

What We Have Covered...

In Chapter 4, we saw chemical "recipes" in action for the first time. We looked at the five most common types of chemical reactions, and practiced "working the numbers" to get the recipe amounts correct through proper balancing. We could imagine setting up the reaction to go, with all of the reactant chemical in their proper amounts, and then we end up with the correct amounts of the correct products. We also saw that these chemical recipes can easily be adjusted through doubling, tripling, halving, and so on by simply dividing through the recipe coefficients with the same number.

But since together we have really only worked these recipes on paper, and some people catch on sooner than later, we might have, at some point, ascribed some degree of speed to the reaction. Those that can identify and balance reactions quickly may be assuming that all reactions proceed from ingredients to products at a similar speed. Others of us, since it takes us a little longer to get the reaction balanced, might imagine that chemical reactions tend to be very slow.

This, I imagine, is not unusual.

But, here are some dots that we may not have connected yet when it comes to chemical reaction "happening":

1. Many chemical reactions occur over *very long periods of time*. For example, the chemical reactions that transform once living organic matter (plants, plankton, animal, etc.) take millions of years to transform into products such as coal ($C_{135}H_{96}O_9NS$), oil, and natural gas (CH_4).
2. Other chemical reactions *occur so fast* that even the fastest digital camera is unable to capture the process "in the act" of occurring.
3. Despite all of the practice problems I have given you, I am sure there must be some that *never* happen (although we can still practice balancing them, just in case).

And finally,

4. During the course of all chemical reactions, there exists a point in time (call it "halftime"), where all that exists is scrambled alphabet soup.
5. There is, always has been, and always will be a heretofore unidentified reactant *and* product. Namely, heat energy. Heat energy is the currency, money, exchange that is *always* given and taken, in different amounts in *every* chemical reaction.
6. We just have not dealt with it yet ... Here we go!

Back in Chapter 3, we discovered a couple of ways to draw fairly complex covalent molecules using either the Chart Method or by playing Electron-MatchMaker. Whichever way you are comfortable with makes no difference. What we need to recognize, again, is that exercise allows us to be the connectivity between atoms in a compound through bonds, whether they are single, double, or triple bonds.

The covalent bonds, formed by two atoms sharing valence electrons, are quite strong and take some effort to break. But broken, they must be! And in order to break *all* of the bonds in *all* of the reactant molecules, a bit of elbow-grease is required, and thus, is the reason that any chemical reaction requires the addition of *input energy* on its reactant side.

It is on the product side of the chemical reaction where we can observe a type of "refund" energy, in the form of output energy. This is the energy that is released as new bonds are formed in the product compounds.

The key concept to remember here, in our discussion of this heat energy budget (also known as enthalpy) is that *never* will the amount of input energy be the same as the amount of output energy. One, or the other, will always be larger.

If there is more input energy than output energy, the reaction is termed endothermic. These endothermic reactions *tend* to be nonspontaneous. They can, and do, occur, but because they require more input energy, they must be continuously "fed" with energy, otherwise they will stop.

If there is more output energy than input energy, the reaction is termed exothermic. These exothermic reactions *tend* to be spontaneous. In other words, once enough input energy is added, the reaction begins to release output energy in a greater amount than the input energy.

(For the Curious: If, by chance, the two energy amounts (input and output) *were* equal, then, the reaction would appear to not even occur! The reactants would change…Back *into* themselves!!)

However, there are certain exceptions to the above tendencies, and to make a more reliable prediction about whether or not the reaction will be spontaneous or nonspontaneous, we must look at one remaining factor, *entropy*.

> **Entropy:** Similar to our discussion of charge earlier in Chapter 2, the meaning of entropy can be somewhat confusing at first. Simply put, entropy is defined as the degree of randomness (or messiness, or sloppiness, or chaos!) that a particular system exhibits. (It is *not* a measurement or quantification of neatness or of order. Just the opposite, in fact!).

A real-life example may be useful here:

> Two science instructors share an office at BPCC, and the office space is shared somewhat equally, on either side of the room. One side of the office, that belonging to Professor Felix, let us say, is very tidy and organized. Every book on this instructor's bookshelf is arranged alphabetically, by publication date, and by descending size. The few items on this instructor's desk are in their proper, expected place, and when moved at all, are quickly returned to their proper spot.
>
> The other side of the office, Professor Oscar's side, is only recognizable as an office thanks to the glow of the computer screen, still just barely visible behind a cascading, cluttered stack of old exams and lab reports.

Kinetic Profiles of Chemical Reactions

Factors Affecting Reaction Kinetics

1.

2.

3.

4.

5.

CHEM 107

Homework #1

Dimensional Analysis

Convert the following measurements using **Dimensional Analysis**. Be sure to:

- Show all work.
- Use valid equalities to form useful conversion factors.
- Cancel unwanted units as you proceed toward your answer.

85.2 inches = ________________ meters

560 grams = ________________ pounds

829 gallons = ________________ milliliters

$2.5 \times 10^{-3} \dfrac{\text{miles}}{\text{second}}$ = ________________ $\dfrac{\text{centimeters}}{\text{hour}}$

The density of helium gas is $0.178 \dfrac{\text{grams}}{\text{Liter}}$. Convert this value to units of $\dfrac{\text{pounds}}{\text{quart}}$.

The air pressure at the top of Mt. Everest has been measured to be, on average, 30,000 Pascals. Express this pressure in units of $\dfrac{\text{lb}}{\text{in}^2}$ (pounds per square inch)

1.09 liters = ________fluid ounces

2130 days = _________seconds

2.3×10^6 millimeters = __________miles

840 miles/hour =_______ meters/hour

If a basketball player is 6 feet, 11 inches tall, how tall is he in millimeters?

The Space Shuttle (when in orbit) traveled at an average velocity of 8,000 meters/second. What is this velocity in "miles/hour"?

CHEM 107

Homework #3

Shorthand Electron Configurations

Write out the entire shorthand electron configuration for the following atoms (or ions)

1. Calcium (Ca)
2. Phosphorus (P)
3. Chromium (Cr)
4. Arsenic (As)
5. Sulfide ion (S^{-2})
6. Aluminum ion (Al^{+3})
7. Zinc (Zn)
8. Ruthenium ion (Ru^{+3})
9. Cadmium (Cd)
10. Manganese ion (Mn^{+7})
11. Bromide ion (Br^{-1})
12. Tin (Sn)

Intro to Ionic Compounds

Homework #4

Ionic Compounds (Formula Writing and Naming)

Part I—For each of the following Ionic Compounds, give the proper name *or* proper formula. (Use your Periodic Table to identify the proper charge for each element!)

1. MgO ______________________
2. sodium fluoride ______________________
3. calcium phosphide ______________________
4. K_3N ______________________
5. $AlBr_3$ ______________________
6. Be_3P_2 ______________________
7. rubidium sulfide ______________________
8. ruthenium oxide ______________________
9. Li_2S ______________________
10. silver phosphide ______________________
11. aluminum nitride ______________________
12. beryllium oxide ______________________
13. CsBr ______________________
14. zinc nitride ______________________
15. zirconium sulfide ______________________

Part II—For each of the following Ionic Compounds, give the proper name or proper formula. (Use your Periodic Table and/or your Polyatomic Ion sheet to identify the proper charge for each element!)

1. chromium (II) phosphide ______________________________
2. $GaPO_3$ ______________________________
3. platinum (IV) sulfide ______________________________
4. $Mo(PO_4)_2$ ______________________________
5. vanadium (IV) chromate ______________________________
6. $Ru(ClO_4)_3$ ______________________________
7. $Ti_3(AsO_4)_4$ ______________________________
8. palladium (II) nitride ______________________________
9. $Ru(NO_2)_3$ ______________________________
10. copper (II) permanganate ______________________________
11. cadmium fluoride ______________________________
12. Au_3AsO_4 ______________________________
13. scandium chloride ______________________________
14. vanadium (III) sulfite ______________________________
15. Rb_2SO_3 ______________________________
16. $Cu(C_2H_3O_2)_2$ ______________________________
17. lithium hydroxide ______________________________
18. Rb_2Se ______________________________
19. osmium (III) iodate ______________________________
20. $Al_2(SO_4)_3$ ______________________________

CHEM 107

Homework #5

Simple (Binary) Covalent Compounds: Formula Writing and Naming

Give the correct **chemical formula** for the following binary **covalent** compounds.

1. trinitrogen pentachloride ______________
2. tricarbon octahydride ________________
3. disilicon nonnitride ___________________
4. hexaphosphorus diselenide __________________
5. pentasulfur heptoxide ___________________
6. mononitrogen trifluoride ____________
7. diphosphorus tetrabromide _____________
8. dihydrogen monoxide _____________
9. hexasulfur octafluoride _____________
10. pentaboron noniodide _______________
11. monocarbon tetrahydride _______________
12. tetranitrogen heptoxide ______________

Give the correct chemical name for the following binary covalent compounds.

13. B_2Br_4 _______________________________________
14. P_4O_8 ____
15. SF_6 _____________________________________
16. Se_5O_{10} _____________________________________
17. N_3S_8 ____________________________________

18. S_8Cl_{10} ______________________________

19. Si_6F_9 ______________________________

20. P_3Cl_6 ______________________________

CHEM 107

Homework #6

Structures of Covalent Molecules

Using the Chart Method, draw an acceptable structure for the following covalent molecules.
Be sure to:

- Satisfy bonding requirements for each atom
- Correctly place lone-pair electrons

a. H_2O_2

b. N_2H_4

c. C_2HBr_3

d. CH_3NH_2

e. HOCN

f. C_3H_6S

g. HO_2N

h. C_3H_4

i. C_2H_2O

j. NSBr

k. HN_2OF

l. SiO_2 (glass)

m. CO_2

n. $C_2H_3Br_3$

o. PBrO

p. N_2HF

q. $C_2O_2H_4$

r. PSiBr

CHEM 107

Homework #9

Double Replacement Reactions

Using the double replacement (ionic "square-dancing") method, correctly predict the products of the reactions below, and balance each reaction correctly.

1. ___$Na_2CO_{3(aq)}$ + ___$HCl_{(aq)}$ →

2. ___$KOH_{(aq)}$ + ___$HCL_{(aq)}$ →

3. ___$NaCl_{(aq)}$ + ___$Na_2CO_{3(aq)}$ →

4. ___$Pb(NO_3)_{2(aq)}$ + ___$NaI_{(aq)}$ →

5. ___$Na_2SO_{4(aq)}$ + ___$BaCl_{2(aq)}$ →

6. ___$Na_2CO_{3(aq)}$ + ___$CaCl_{2(aq)}$ →

7. ___$KOH_{(aq)}$ + ___$Mg(NO_3)_{2(aq)}$ →

8. ___$Na_2SO_{4(aq)}$ + ___$Cu(NO_3)_{2(aq)}$ →

9. ___$HCl_{(aq)}$ + ___$KHCO_{3(aq)}$ →

10. ___$Sr(NO_3)_{2(aq)}$ + ___$KIO_{3(aq)}$ →

CHEM 107

Homework #10

Stoichiometry Practice Problems

- Be sure to correctly balance the reactions.
- Use your Periodic Table for any atomic weights needed.

1. A chemist performs the reaction below, starting with a total of 8.27 grams of aluminum. How many grams of aluminum bromide ($AlBr_3$) should she expect to produce?

$$\_\_\_HBr + \_\_\_Al \rightarrow \_\_\_AlBr_3 + \_\_\_H_2$$

2. A bunsen burner uses methane (CH_4) as its fuel. According to the reaction below, how many grams of water (H_2O) would be produced if 0.085 grams of methane reacts?

$$\_\_\_CH_4 + \_\_\_O_2 \rightarrow \_\_\_CO_2 + \_\_\_H_2O$$

If 37 grams of calcium nitrate are used as a reactant in the following reaction, how many grams of potassium nitrate should (theoretically) be produced?

$$\_\_\_\_Ca(NO_3)_2 + \_\_\_\_KOH \rightarrow \_\_\_\_Ca(OH)_2 + \_\_\_\_KNO_3$$

If a chemist runs the reaction described above, but only produces 31.9 grams of potassium nitrate, what is the Percent Yield for this reaction?

If 214 grams of aluminum hydroxide are used as a reactant in the following reaction, how many grams of water should (theoretically) be produced?

$$\_\_\_\_Al(OH)_3 + \_\_\_\_H_3PO_4 \rightarrow \_\_\_\_AlPO_4 + \_\_\_\_H_2O$$

If a chemist runs the reaction described above, but only produces 99.2 grams of water, what is the Percent Yield for this reaction?

If 0.26 grams of tetraphosphorus decoxide are used as a reactant in the following reaction, how many grams of phosphoric acid (H_3PO_4) should (theoretically) be produced?

$$\_\_\_\_P_4O_{10} + \_\_\_\_H_2O \rightarrow \_\_\_\_H_3PO_4$$

If a chemist runs the reaction described above, but only produces 0.175 grams of phosphoric acid, what is the Percent Yield for this reaction?

CHEM 107

Homework #14

Strong Acid/Base Neutralization Reactions

Skills and reminders:

- Acid/strong base neutralization reactions are double replacement reactions ($AB + CD \rightarrow AD + CB$)
- Water (H_2O or HOH) and a salt are *always* the products.
- Writing ionic compound formulas (±)
- Balancing reactions
- Correctly naming products

Correctly predict the products of the following acid/base neutralization reactions, and balance the reaction, as needed.

1. ___$H_2SO_{4(aq)}$ + ___$Ca(OH)_{2(aq)}$ $\rightarrow$
2. ___$Al(OH)_{3(aq)}$ + ___$HBr_{(aq)}$ $\rightarrow$
3. ___$H_3PO_{4(aq)}$ + ___$NaOH_{(aq)}$ $\rightarrow$
4. ___$Fe(OH)_{2(aq)}$ + ___$HF_{(aq)}$ $\rightarrow$
5. ___$HNO_{3(aq)}$ + ___ $Mg(OH)_{2(aq)}$ $\rightarrow$
6. ___$H_2SO_{3(aq)}$ + ___$LiOH_{(aq)}$ $\rightarrow$

Weak Acid/Base Reactions (Buffer Reactions)

Weak acid/weak base reactions are simply "proton ping-pong."

- Weak acid (and conjugate acid) *lose* their H^+
- Weak base (and conjugate base) *gain* an H^+
- Water, being amphiprotic, will play the role of the acid or the base, as needed.
- Be able to label the:

  Weak acid

  Weak base

  Conjugate acid

  Conjugate base

  $H_2O_{(l)} + NH_{3(aq)} \longleftrightarrow$

  $HC_2H_3O_{2(aq)} + H_2O_{(l)} \longleftrightarrow$

  $H_2O_{(l)} + HCO_2H_{(aq)} \longleftrightarrow$

  $NBr_2F_{(aq)} + H_2O_{(l)} \longleftrightarrow$

  $H_2O_{(l)} + HCN_{(aq)} \longleftrightarrow$

Chart Method for Covalent Molecules

Part I—CHART PREPARATION

1. "Smash" the molecule into its atoms (e.g., CH_3Br becomes C H H H Br) and write them along top of chart.
2. Underneath the atoms, write the *number of valence electrons* for each. (Check the *Group #* of each element on Periodic Table.) Add these numbers to get the total valence electrons (TVE).
3. Next, write the *number of bonds each atom needs*. (This can be found by ensuring each column adds up to "8," except for Hydrogen which will only add to "2.")

Part II—STRUCTURE OF MOLECULE

4. Choose the "Central Atom." This is the atom with the *greatest number of bonds needed* (from Step 3). Draw the central atom in the center of page. [If there is a "tie," draw both, and connect with a single line (single bond).]
5. Using only single bonds (–), connect each peripheral atom, one at a time, Subtract *two electrons* from the TVE for *each bond* that you draw.
6. Ensure that *each atom* in your molecule drawing has the *correct number of bonds* (check back with Step 3, above).

Part III—"LONE PAIR" ELECTRON PLACEMENT

7. Place remaining TVE electrons around the peripheral atoms (in pairs) to ensure that all peripheral atoms can claim "8" valence electrons. (Remember, *each bond* "equals" *two electrons* that can be shared by both atoms.) Subtract any electrons you use from the TVE.
8. Add any remaining electrons to the central atom, if it needs them, to ensure it has "8" valence electrons.
9. You *must* finish with exactly "0" electrons!!

Hints: **a.** Carbon will *always* have four bonds, and *no lone pair electrons*. It is the most common *central* atom.
b. Hydrogen will always have *one, and only one*, bond, and *no* lone pair electrons.
c. Nitrogen and phosphorus, as central atoms, will have a lone pair of electrons placed on them (Step 8).

The Five Common Types of Chemical Reactions

(there are other, less-common, types as well)

1. **Synthesis (Combination) Reactions:** Two or more reactants combining to produce a single product.

   General Form: A + B + C $\rightarrow$ ABC

2. **Decomposition Reactions:** One reactant decomposing to produce two (or more) products.

   General Form: ABC $\rightarrow$ A + B + C

3. **Combustion Reactions:** A hydrocarbon (or carbohydrate) reacts as a *fuel* with oxygen (O_2) to produce carbon dioxide (CO_2) and water (H_2O).

   General Form: $C_xH_y + O_2 \rightarrow CO_2 + H_2O$

4. **Single Replacement Reactions:** A single element (C, N_2, Si, Mg, O_2, etc.) reacting with a binary compound to produce another single element and a different binary compound.

   General Form: AB + C $\rightarrow$ BC + A

5. **Double Replacement Reactions:** Two ionic compounds react to produce two new ionic compounds (and sometimes H_2O). (Ionic "Square-Dancing.")

   General Form: AB + CD $\rightarrow$ AD + CB

CPSIA information can be obtained
at www.ICGtesting.com
Printed in the USA
LVHW060254240419
615166LV00002B/18/P

9 781465 275813